I0082410

Today's Forecast— Tomorrow's Weather?

Weather, Wisdom, and Life Under Northwest Skies

Patrick Timm

Coastal Tides Press

Copyright © 2026 by Patrick Timm

All rights reserved.

No portion of this book may be reproduced in any form without written permission from the publisher or author, except as permitted by U.S. copyright law.

Coastal Tides Press

ISBN: **979-8243209502**

Contents

Introduction

Everyone talks about the weather. Nobody does anything about it. Which is probably for the best.

Weather is the one constant we all share and none of us control. It arrives whether we're ready or not, shaping our days, our moods, and sometimes our memories. In the Pacific Northwest, it does this with particular subtlety. Our weather rarely shouts. More often, it nudges, lingers, or quietly changes the tone of a day.

After watching the Northwest sky for more than six decades, I've come to believe there's no such thing as bad weather—only different kinds of good weather. Some invite us outside. Others ask us to slow down, pay attention, or stay close to home. All of them have something to teach us, if we're willing to look up.

This book grew out of that habit of watching. Years ago, I came across a modest, thoughtful volume published in 1978—*Weather of the Pacific Coast* by Walter D. Rue. It was written in a different era, before modern satellite imagery and computer models filled our screens. Rue didn't try to predict the weather so much as explain what usually happens, why it happens, and how people along the coast had learned to live with it.

This book is written in that same spirit, but from a different time. It draws on decades of observation, modern understanding, and a long-running conversation with readers who care deeply about what the sky is doing—and what it means for everyday life in the Northwest.

What follows is not a forecast. It's a yearlong walk through Pacific Northwest weather, month by month, with a focus on Seattle and Portland. Along the way, you'll find old weather sayings, simple explanations, familiar patterns, and the occasional surprise. You'll also find numbers—averages, not promises—because weather has always had a way of reminding us that it gets a vote.

Today's forecast may hint at tomorrow's weather. But the atmosphere, like life, always reserves the right to change its mind.

January

Quiet Cold

January in the Pacific Northwest rarely announces itself. The holidays fade, the lights come down, and the days begin to lengthen by minutes rather than hours. Winter doesn't arrive with drama here. It settles in quietly.

This is often the coldest-feeling month of the year, even if the thermometer doesn't always show it. Clear nights allow heat to escape, and mornings can begin with frost on rooftops and windshields. When storms do arrive, they tend to be steady rather than spectacular—rain that lingers, clouds that hang on, and the occasional surprise when cold air sneaks in at just the right moment.

Snow is never a guarantee in January, but it's always a possibility. When it comes, it often arrives with little warning and leaves just as quickly. More often, January teaches patience. It's a month for watching the sky, noticing the light, and understanding that winter here speaks softly.

January also marks a subtle emotional shift. December carries expectation—holidays, gatherings, lights against the dark. January is quieter and more honest. The decorations are gone, routines return, and the season shows itself without distraction. In the Northwest, this often means low clouds, damp afternoons, and mornings that take their time warming up.

It's also a month when weather myths begin to surface. A sunny January day can feel misleading, giving the impression that winter is easing its grip. In reality, these clear stretches often accompany the coldest air of the season. January reminds us that appearances can be deceiving, especially when it comes to weather.

Seattle & Portland

January behaves a little differently depending on where you live. Seattle often holds onto cloud cover longer, while Portland is more vulnerable to sharp cold when east winds push through the Columbia River Gorge. They are caused by higher pressure over the Columbia Basin and ushers in cold arctic air from the north. Wind chills can be extreme. Once the cold air settles in over the Portland-Vancouver metro area, residents watch for overrunning of milder Pacific air resulting in sometimes heavy snow and ice. Those winds can turn an otherwise calm winter day into one that feels much colder than expected.

The warmer waters of Puget Sound keep Seattle proper relatively mild compared to outlying suburbs to the east. Snowfall is brief unless the Puget Sound Convergence Zone develops. This usually occurs in Everett to the north as frigid cold moves south out of the Fraser River Canyon near Bellingham and slowly sags southward toward the Sound. This is Seattle's access to the arctic air.

Longtime residents learn to watch the pressure patterns as much as the clouds. East wind events, while not constant, define January's cold snaps in the Portland area. In Seattle, cold air more often sneaks in from the north, filtered by water and terrain. These differences explain why one city may see freezing fog while the other remains merely damp and gray.

January in the Pacific Northwest favors short outings and familiar routes, timed between rain and frost.

In Seattle, waterfront walks remain popular, especially around Elliott Bay and Alki Beach, where open views and moving air make even gray days feel expansive. Locals take advantage of dry breaks for brisk walks, knowing cafés and warm indoor spaces are never far away. Urban parks and neighborhood loops are favored over long hikes, particularly during frosty mornings.

In Portland, residents continue to use neighborhood trails and city parks throughout January. Forest Park sees steady foot traffic on lower, well-drained trails, while riverside paths attract walkers bundled against the chill. January is also a favorite month for birdwatching, as winter species are active and foliage is minimal, improving visibility.

January outings tend to be unhurried and purposeful — a walk for fresh air, a brief visit outdoors, then back to warmth.

January dressing in both Seattle and Portland is about staying dry, warm, and flexible rather than bracing for extremes.

In Seattle, damp air and persistent cloud cover can make temperatures feel colder than they are. A waterproof jacket with a hood is essential, as rain is often light but frequent. Layers work best—sweaters or fleece under a shell—since indoor spaces are usually well heated. Comfortable, water-resistant shoes are a must for slick sidewalks. Umbrellas are optional and often unused; most locals rely on hoods and habit.

In Portland, winter attire follows a similar pattern but with added attention to cold snaps driven by east winds through the Columbia Gorge. These winds can make otherwise calm days feel sharply colder. A warm, wind-resistant jacket is helpful during these periods. Waterproof footwear is important, especially when temperatures hover near freezing. Umbrellas are widely used and perfectly acceptable here.

Winter dressing in the Northwest favors readiness over routine—layers allow you to adapt as the day unfolds.

If you like numbers, here's what January usually looks like.

Seattle
- Average high: 47.2 degrees F

- Average low: 34.1 degrees F

- Average mean temperature: 42.3 degrees F

- Average monthly rainfall: 5.14 inches

Extremes:
- Maximum High: 67 degrees F, 1931

- Minimum Low: 11 degrees F, 1950

- Precipitation: 10.93 inches, 1953

Portland
- Average high: 48.4 degrees F

- Average low: 34.3 degrees F

- Average mean temperature: 41.9 degrees F

- Average monthly rainfall: 5.00 inches

Extremes:
- Maximum high: 66 degrees F, 2003

- Minimum low: -2 degrees F, 1950

- Precipitation: 12.83 inches, 1953

Snowfall varies widely from year to year. Some months of January pass quietly, while others leave a brief but lasting impression.

WEATHER LORE

"As the days lengthen, the cold strengthens."

There's more truth in this old saying than many people expect. As daylight slowly returns, clear skies often allow temperatures to fall at night. January reminds us that longer days don't immediately mean warmer ones. Winter isn't finished yet—but the slow turn toward spring has already begun.

"A January thaw never bodes well."

January thaws have long made people uneasy, and for good reason. In the Pacific Northwest, a brief warm spell in midwinter often signals a shift in weather rather than a lasting change. Mild air can bring heavy rain, rising rivers, and saturated ground, sometimes followed by a return to colder conditions. January teaches that warmth this time of year is usually temporary—and often comes with complications.

"Frost in January brings good luck."

This saying likely comes from generations who welcomed frost as a sign of clear skies and stable conditions. In the Northwest, frost often arrives during calm, high-pressure weather—the kind that brings bright mornings and crisp air. While frost doesn't promise anything about the months ahead, it does offer a reminder that winter has its own quieter kind of beauty.

FROM THE WEATHER EYE

January cold has a way of sneaking in under blue skies. These are the days when the air feels sharper than it looks, and the quiet itself seems chilled. It's the kind of cold that doesn't rush you—it asks you to notice, to slow down, to understand that winter in the Northwest isn't about extremes, but about quiet shifts that shape the rhythm of the season.

Because snowfall is uncommon, memorable storms tend to loom large in Northwest weather history. When snow does arrive in January, it often does so under just the right combination of cold air and incoming Pacific moisture. These setups are rare, but when they occur, they leave an impression that lasts for decades.

Although snowfall in Portland and Seattle is usually light in January, there have been notable exceptions. In Portland, January 1950 brought a record 41 inches of snow. In January 1980, roughly 20 inches fell over several days. More recently, a fast-moving storm in January 2017, overriding chilly east winds, delivered a foot or more of snow across many areas. It's often said that in Portland, it takes only an inch of snow to bring traffic to a stand-still—and history has supported that claim more than once.

Seattle has had its share of memorable January snow-storms as well. Records tell of 1880, when snow piled several feet deep in downtown Seattle. On January 13, 1950, 20 inches fell in a single day—still remembered as the Great Snowstorm. In 1979, 18 inches accumulated over three days, and some residents recall January 2012 as another snowy chapter. These events happen, but rarely. Many Januarys pass without snow at all.

These storms stand out precisely because they are the exception. Most Januarys are quiet, reinforcing the under-stated nature of winter west of the Cascades.

I reflect on January as a midwinter month that carries both darkness and light. Brief spells of mild weather coax daffodils to peek upward, as if spring can't be far away.

A few bold robins may appear in backyard shrubs. Pussy willows usually save their entrance for February, but now and then they arrive early. Keep watching the sky—and your surroundings. Nature may be reserved this month, but it still manages a smile once in a while.

QUICK TAKEAWAY

January at a glance
- Cold often feels sharper under clear skies

- Snow is possible, but never promised

- East winds can make calm days feel colder

- Longer days don't mean warmer weather—yet

February

Holding Pattern

February is a month that tests patience in the Pacific Northwest. Winter hasn't loosened its grip, but the promise of spring feels closer now, even if it's mostly imagined. Daylight increases more noticeably, afternoons linger a little longer, and the sky occasionally hints at change—only to retreat again into gray.

This is the heart of midwinter. Storms still arrive, rain remains frequent, and cold air can reassert itself without much warning. February often feels repetitive, as if the atmosphere is circling the same idea rather than moving on to something new. Yet beneath that sameness, subtle shifts are underway.

High-pressure systems occasionally bring dry, crisp days that feel almost springlike, especially when the sun breaks through. It is not unusual to have several days mid-month reach well into the 60s. These stretches rarely last, but they're enough to remind people that winter does have an end. February doesn't rush the season—it holds it in place, waiting.

Seattle & Portland

February weather continues to reflect the differences between Seattle and Portland, shaped by geography and airflow. Seattle often remains cloudier, with steady light rain and damp conditions that linger day after day. This can make the city feel dark and the dampness penetrates to the bone. A good month for indoor activities. Portland, while similar in many ways, is more susceptible to temperature swings tied to east winds through the Columbia Gorge. As high pressure slides east of the mountains, dry, cool winds rush out of the Gorge. Sometimes they actually help warm the atmosphere giving one a sense of a false spring.

Cold air outbreaks remain possible in both cities, though they tend to be shorter-lived than those in January. When cold air does arrive, it often sets the stage for freezing rain or mixed precipitation, especially in the Portland-Vancouver area. Snow remains possible, but it's rarely persistent.

February clothing mirrors January but with slightly more flexibility as brief mild spells begin to appear.

February feels slightly more hopeful, even when winter weather lingers. Locals watch closely for brief mild spells, and when they arrive, they don't waste them.

In Seattle, longer daylight encourages afternoon walks around Lake Union and neighborhood greenways. When skies brighten, residents head outside quickly, knowing the window may be short. Waterfront viewpoints remain popular, especially on clearer days when distant mountains briefly reveal themselves.

In Portland, February often brings a few surprisingly mild stretches. These days draw people back to the waterfront, neighborhood cafés with outdoor seating, and local parks. Early gardening tasks begin during dry spells, even if planting is still weeks away. Frog choruses and early bird activity signal slight seasonal change, rewarding those who step outside and listen.

February outdoor life is about paying attention — not pushing the season, but noticing its early shifts.

In Seattle, layers remain the key. Rain is still common, but dry stretches occur often enough that lighter jackets may come out on sunnier days. A waterproof shell is still essential, but heavy winter coats are less consistently needed. Shoes should handle moisture, and sunglasses occasionally make a welcome reappearance during bright afternoons.

In Portland, February often delivers a few surprisingly mild days, sometimes allowing lighter jackets or sweatshirts to replace heavier winter wear. That said, winter isn't finished, and cold air can return quickly. Layers are essential, with a waterproof jacket close at hand. As in January, umbrellas are common and practical. Footwear should still be chosen with ice or wet pavement in mind.

Winter dressing in the Northwest favors readiness over routine—layers allow you to adapt as the day unfolds.

If you like numbers, here's what February usually looks like.

Seattle

- Average high: 49.8 degrees F

- Average low: 37.0 degrees F

- Average mean temperature: 43.4 degrees

- Average monthly rainfall: 3.54 inches

Extremes:

- Maximum high: 74 degrees F, 1968

- Minimum low: 11 degrees F, 1989

- Precipitation: 8.16 inches, 2017

Portland

- Average high: 51.5 degrees F

- Average low: 36.8 degrees F

- Average mean temperature: 44.1 degrees F

- Average monthly rainfall: 3.68 inches

Extremes:

- Maximum high: 71 degrees F, 1988

- Minimum low: minus 3 degrees F, 1950

- Precipitation: 10.36 inches, 2017

February is quieter statistically than January, but it still has a way of surprising people—often at inconvenient moments.

WEATHER LORE

"If February gives much snow, a fine summer it doth foreshow."

Sayings like this reflect a long-standing belief that winter sets the tone for what follows. While modern forecasting doesn't support such simple cause-and-effect thinking, there's something comforting in the idea. A cold, snowy February feels productive, as if winter is doing its job thoroughly before stepping aside.

In the Northwest, however, February snowfall is unpredictable. Some years bring none at all. Others deliver a brief burst that quickly melts away. The saying survives not because it's reliable, but because it expresses hope—something February quietly invites.

"February fills the ditches, black or white."

This old saying captures February's unpredictability perfectly. Whether it's rain filling roadside ditches or snow briefly whitening the ground, February rarely stays neutral. In the Pacific Northwest, this often means plenty of water moving through creeks and gutters, sometimes quietly, sometimes all at once. Either way, February tends to make its presence known.

"February brings the rain that wakes the grain."

Farmers once watched February closely, knowing winter moisture helped prepare the ground for spring growth. In the Northwest, February rain often feels relentless, but it plays an important role—recharging soils, swelling streams, and setting the stage for what follows. It may not feel productive at the time, but February weather is quietly doing its work.

"When February comes, winter's grip is half undone."

There's a hopeful tone to this saying, even if reality doesn't always cooperate. February brings longer days and gradual signs of change, but winter still holds the reins. In the Northwest, this often means the season pauses rather than progresses, offering hints of what's coming without making any promises.

FROM THE WEATHER EYE

February often feels like a month caught between decisions. Winter hasn't finished, but spring hasn't committed either. The result is a stretch of weather that repeats itself just enough to feel familiar, sometimes even dull, as if the atmosphere is waiting for a cue.

And yet, this is when small signs begin to stand out. A brighter patch of sky late in the afternoon. The sound of water moving more freely through creeks and gutters. A brief stretch of dry weather that invites people outside, even if coats are still required. After weeks of sameness, these moments feel larger than they are.

When the sun does shine in February, native Northwesterners respond instinctively. Shorts appear. Athletic shoes or sandals replace boots. Shirtsleeves and light sweaters make a cautious return, along with sunglasses that haven't been used in months. On a sunny Sunday afternoon, you might even spot a classic car rolling by with the top down—maybe for show, maybe just for the glow.

Almost without fail, especially in Portland, February seems to offer a short mid-month break. A few mild, dry days arrive unexpectedly, and temperatures can climb well into the 60s. It's often just enough warmth to lift the winter mood that's been hanging around since December, even if everyone knows it won't last.

A brief lull in Pacific storms, combined with a mild night or two, can awaken the tiny frogs that have been waiting quietly through winter. Their chorus, sudden and unmistakable, can be heard from Portland to Seattle and beyond—an early reminder that the season is slowly shifting beneath the surface.

Snow in February tends to be more disruptive than impressive. Roads glaze over, commutes slow, and patience wears thin. These events are usually short-lived, but they leave an impression. More often, February's contribution is quieter—a continuation of winter with just enough variation to keep us paying attention.

As meteorological winter comes to an end, I'm reminded of a line from Sinclair Lewis, who once wrote, *"Winter is not a season, it's an occupation."* Here in the Pacific Northwest, we take it in stride. February may not signal the end of winter's work, but it does suggest that the job is nearly done.

QUICK TAKEAWAY

February at a glance

- Winter holds steady, but daylight increases noticeably

- Storms remain common, snow less so

- Brief dry spells offer hints of what's coming

- Patience is February's most reliable forecast

March

FALSE STARTS

March has a reputation for impatience. After months of gray, rain, and routine, it arrives carrying the promise of change. And sometimes, briefly, it delivers. A warm afternoon. A stretch of blue sky. A hint of spring in the air.

Then it changes its mind.

In the Pacific Northwest, March is a month of mixed signals. It's when winter begins to loosen its grip, but not without resistance. Showers come and go quickly, often in rapid succession. Sunbreaks appear just long enough to tempt people outside before the next cloud slides in. March doesn't settle—it experiments.

Contrast becomes the defining feature now. One day may feel almost mild, the next sharply cool. Snow levels rise and fall, rain turns to hail, and winds shift direction without much notice. March weather keeps you guessing, reminding us that spring is a process, not an event.

Despite its unpredictability, March brings progress. Daylight increases noticeably, evenings linger, and the landscape responds. Lawns green up, buds swell, and early blossoms test the air. March may stumble, backtrack, and hesitate, but it always moves forward—eventually.

Seattle & Portland

March behaves differently depending on where you live, but unpredictability is the common thread. Seattle often experiences frequent showers punctuated by sunbreaks, especially during the afternoon. These passing showers can be light or briefly intense, sometimes producing small hail as colder air aloft moves overhead. It's not unusual to be at Pike Place Market, enjoying a sunbreak and holding a hot cappuccino, while pedestrians on Queen Anne Hill are navigating ice pellets or a brief burst of snow.

Spring begins to announce itself in subtle ways. One of the most reliable signs arrives with the cherry blossoms at the University of Washington, especially around the Quad. Even so, March is not the time to abandon winter habits. Clothing choices here look much like February's—layers remain essential. A sunbreak may offer a brief reprieve from the chill of Puget Sound, only to be followed by quickly falling temperatures once the clouds return.

A lightweight, roll-up raincoat is a good idea. Umbrellas are best left at home unless you don't mind being noticed as a newcomer. Comfortable, water-resistant shoes are far more important.

Portland sees similar variability, though east winds through the Columbia Gorge can still play a role early in the month. Cold air lingering east of the Cascades occasionally slips westward, sharpening temperatures and complicating the transition to spring. When milder Pacific air rides over that cold layer, mixed precipitation is still possible, though it becomes less likely as the month progresses.

Clothing choices in Portland also favor layers, as conditions can change noticeably from hour to hour. Comfortable walking shoes—preferably waterproof—are always a good idea. A waterproof jacket is essential, and unlike Seattle, umbrellas are welcome here. You'll see them everywhere, especially on days when the rain settles in rather than passing through. A mainstay for walking the waterfront.

If you like numbers, here's what March usually looks like.

Seattle

- Average high: 53.9 degrees F

- Average low: 39.2 degrees F

- Average mean temperature: 46.6 degrees

- Average monthly rainfall: 3.86 inches

Extremes:
- Maximum high: 78 degrees F, 2016

- Minimum low: 20 degrees F, 1897

- Precipitation: 8.74 inches, 1997

Portland
- Average high: 56.8 degrees F

- Average low: 39.7 degrees F

- Average mean temperature: 48.3 degrees

- Average monthly rainfall: 3.97 inches

Extremes:
- Maximum high: 82 degrees F, 2025

- Minimum low: 19 degrees F, 1989

- Precipitation: 7.89 inches, 2012

March statistics suggest improvement, but the atmosphere often prefers improvisation.

WEATHER LORE

"March comes in like a lion and goes out like a lamb."

This may be the most quoted weather saying of the year—and one of the least reliable. In the Pacific Northwest, March rarely follows a neat script. It may roar at the beginning, soften at the end, or reverse the order entirely. More often, it behaves like both at once.

The saying survives because it captures how March feels. There's motion, restlessness, and a sense that something is trying to change. The details may vary, but the transition is real.

"A dry March, a wet May."

Sayings like this remind us how much people once watched the calendar—and worried about what came next. In reality, March in the Northwest is rarely truly dry, but the sentiment holds. When March offers more sun than expected, it often feels borrowed rather than earned. The atmosphere has a way of balancing things out, usually on its own schedule.

"March weather is never settled."

This may be less poetic than some sayings, but it's hard to argue with. March resists routine. One day hints at spring, the next reminds us winter still knows the way back. In the Northwest, this restlessness is normal—and useful. March weather keeps us alert, reminding us that seasonal change is rarely neat or immediate.

"March winds and April showers bring forth May flowers."

There's truth tucked inside this old saying, even if the timing isn't always exact. In the Pacific Northwest, March wind and rain often feel like more of the same after winter, but they play an important role. These unsettled weeks mix the atmosphere, redistribute moisture, and help wake a landscape that's been waiting quietly for its cue. March may not feel productive, but it's laying the groundwork.

FROM THE WEATHER EYE

March is the month when people start paying attention again. After weeks of routine winter weather, even a brief sunbreak feels like a reward. Coats get lighter, windows open for a moment, and the outside world begins to call.

Weather-wise, March is rarely dramatic, but it is busy. Showers move through quickly, skies brighten and darken in the same afternoon, and the wind often shifts direction without warning. Hail showers can appear suddenly, rattle briefly, and vanish just as fast. Snow levels rise and fall, sometimes within hours, adding to the sense that nothing has quite settled yet.

These changes can be frustrating, especially for those hoping winter is finished. But March weather serves a purpose. It mixes air masses, redistributes moisture, and prepares the atmosphere for the steadier patterns of spring. The back-and-forth is part of the process.

I think of March as the season learning to walk again. It wobbles, stumbles, and occasionally falls back into old habits. But with each passing day, the light grows stronger, the soil slowly warms, and the signs of spring become harder to ignore.

At local parks, you'll find kites rising high in the sky, competing with towering white cumulus clouds. Early spring training for Little League baseball puts both players and parents to the test, as weather interrupts practice as often as it cooperates.

I remember bouts of snowfall in March, but it's usually wet and sloppy—fairing well on bark dust and grass, but not so much on roadways. After clouds and precipitation pass, the higher angle of the sun quickly melts away the evidence. What snow?

March is a feisty month to predict, much like house-training a new puppy. Take your eyes off it for a moment, and everything changes. It's entirely possible for scattered thundershowers with small hail to pop up unexpectedly, adding to the surprises hidden in passing clouds overhead.

St. Patrick's Day often brings scattered rain showers and bursts of small hail. Local residents keep an eye on the sky, watching for rainbows and half-joking about chasing down that elusive pot of gold.

March 1 marks the beginning of meteorological spring, though many people wait for the astronomical start later in the month. It's that time of year when seasonal boundaries blur, and a little atmospheric chaos is part of the deal.

Weather lore suggests that if frogs holler early in March, three more freezes are on the way. And they do get vocal at times this month. While the average last frost for Portland and Seattle draws closer, April usually holds that distinction. March reminds us that spring may be arriving—but winter hasn't fully packed up yet.

March doesn't demand certainty—only attention.

QUICK TAKEAWAY

March at a glance
- Expect rapid changes, sometimes within the same day

- Showers replace steady rain

- Sunbreaks are frequent but brief

- Spring arrives gradually, not all at once

April

BETWEEN SHOWERS

April arrives with a lighter touch. The days are longer now, the air feels softer, and the grip of winter has noticeably loosened. While rain is still part of the picture, it no longer dominates the way it did earlier in the season. April weather feels transitional, but with more confidence than March ever managed.

This is the month when showers replace storms. Rain tends to arrive in short bursts rather than long stretches, often followed by brighter skies and occasional sunbreaks. The atmosphere seems more willing to move along, and the landscape responds quickly. Lawns green up, trees leaf out, and spring begins to feel less like a promise and more like a presence.

April is also a month of contrast, but the swings are gentler now. Cool mornings give way to mild afternoons, and evenings linger comfortably longer. It's the kind of weather that encourages people outside—sometimes without much planning—because the odds feel better than they have in months.

Seattle & Portland

April behaves similarly in Seattle and Portland, though local influences still shape the details. Seattle often sees a steady mix of light showers and sunbreaks, with rainfall spread out across the month rather than concentrated in long events. When the sun appears, even briefly, it tends to feel warmer now, thanks to the higher spring sun angle.

Portland experiences comparable conditions, with slightly warmer afternoons becoming more common as April progresses. East winds through the Columbia Gorge are far less frequent, and cold air intrusions become increasingly rare. Rainfall remains part of daily life, but it's often lighter and easier to work around.

April is when the Pacific Northwest begins to stretch its legs. The days are longer now, showers are lighter and more scattered, and outdoor plans feel more attainable—even if they still require flexibility.

In Seattle, waterfront walks grow noticeably busier. Paths along Elliott Bay, Alki, and Lake Union invite strolling during sunbreaks, with easy shelter nearby when showers pass through. April is also peak season for cherry blossoms throughout the city. Pike Place Market remains lively, offering an ideal mix of open-air exploration and indoor refuge.

In Portland, April brings renewed energy to parks and neighborhoods. Forest Park trails see more hikers, as spring growth fills in beneath the canopy and muddy stretches begin to firm up. The Willamette River waterfront draws walkers, runners, and cyclists whenever skies brighten. Farmers markets expand their offerings, signaling that the growing season is truly underway.

In both cities, April also marks the return of outdoor routines—youth sports resume more consistently, gardeners begin planting in earnest, and open parks fill with kites, frisbees, and the steady optimism that winter is finally loosening its grip.

April dressing favors adaptability over insulation.

In both Seattle and Portland, a lightweight waterproof jacket remains essential, but heavier winter coats are usually left behind. Layers are still important—cool mornings often give way to mild afternoons, especially during sunbreaks.

Comfortable water-resistant walking shoes are recommended, as trails and sidewalks can remain damp even after rainfall ends. In Portland, umbrellas are common and practical; in Seattle, most people continue to rely on hoods and timing their outings between showers.

Sunglasses become more useful in April, as sunbreaks are brighter and more frequent. A light sweater or fleece layer is often all that's needed once clouds part.

April's rule of thumb is simple: dress for the morning, adjust for the afternoon, and enjoy the growing stretches of spring in between.

If you like numbers, here's what April usually looks like.

Seattle
- Average high: 58.8 degrees F

- Average low: 42.8 degrees F

- Average mean temperature: 50.8 degrees F

- Average monthly rainfall: 2.98 inches

Extremes:
- Maximum temperature: 89 degrees F, 2016

- Minimum low temperature: 30 degrees F, 1920

- Precipitation: 5.80 inches, 1996

Portland

- Average high: 62 degrees F

- Average low: 43.7 degrees F

- Average mean temperature: 52.8 degrees F

- Average monthly rainfall: 2.89 inches

Extremes:

- Maximum temperature: 90 degrees F, 1998

- Minimum temperature: 29 degrees F, 1955

- Precipitation: 5.73 inches, 2022

April's numbers reflect improvement, even if umbrellas are still within reach.

WEATHER LORE

"April showers bring May flowers."

This saying has survived for a reason. In the Pacific Northwest, April rain often arrives just when the landscape needs it most. Showers are frequent enough to support growth but usually light enough to allow time outdoors in between. April rain doesn't linger—it nourishes.

While the timing may vary, the message holds. April weather is productive, even when it feels inconvenient. It's the month when the groundwork for late spring is quietly laid.

"If it thunders in April, frost in May."

Thunderstorms are uncommon west of the Cascades, but April can surprise us with brief bursts of thunder, small hail, or even lightning. These events often signal colder air aloft, and while the saying isn't a guarantee, it hints at April's unsettled nature. Warm afternoons can still give way to chilly nights. It's a good reminder not to rush tender plants outdoors just yet. "

April wind shakes the ice from the pond."

April winds are often overlooked, but they play an important role. Gusty afternoons help mix the atmosphere, clearing lingering cold air and drying saturated ground between showers. While the wind can feel sharp at times, it's part of the transition—stirring the season forward and nudging winter toward the exit.

"April weather—rain in the morning, sunshine by noon."

This saying feels tailor-made for the Pacific Northwest. April often delivers a little of everything in a single day, encouraging flexibility rather than firm plans. Morning showers give way to brighter afternoons, and optimism rises quickly once the clouds part. It's the month that teaches locals to wait an hour before deciding whether the day is a loss—or a gift.

FROM THE WEATHER EYE

April is when the Northwest starts to look awake again. After months of muted color, the return of green feels sudden and almost exaggerated. Trees leaf out quickly, gardens come to life, and neighborhoods seem to expand outward as people spend more time outside.

Weather-wise, April rewards flexibility. A passing shower may interrupt plans briefly, but it rarely ends them. Sunbreaks appear more often and last longer, warming sidewalks and storefronts just enough to encourage lingering.

There's also a noticeable change in how the rain feels. April showers are lighter, less insistent, and easier to tolerate. They arrive, pass through, and move on, leaving behind fresh air and a sense that something has been accomplished.

I think of April as the month when patience pays off. Winter's long occupation finally winds down, replaced by a season that feels willing to cooperate. The weather may still interrupt, but it no longer resists. Spring has found its footing.

Over the years of writing my column, I've also seen April remind us that it's not without its edge. In Portland, following a cold frontal passage, strong instability can develop when cold air aloft overrides warming surface temperatures. This setup can produce vigorous showers, hail, and at times severe thunderstorms.

April is the prime month for what meteorologists call cold-core funnel clouds. These narrow funnels often form beneath towering spring clouds, usually remaining aloft and causing little concern. Occasionally, however, one will briefly touch down, producing a weak tornado with minimal damage. Similar setups can occur again in October, but April remains the most active period.

One April event stands apart. On April 5, 1972, a powerful tornado developed over the Columbia River near Portland and moved north into Vancouver, Washington. Unlike most Northwest tornadoes, this one caused significant destruction. Six people were killed, and millions of dollars in damage were left behind. Classified as an F3 tornado, it remains the strongest and most destructive tornado in Washington State history.

April usually greets us with gentle rain and fresh growth, but it also carries reminders of the atmosphere's complexity. It's a month that invites optimism—while quietly asking for respect.

In Puget Sound, tornadoes and funnel clouds are uncommon, thanks to the region's complex geography and the moderating influence of relatively warm water. When they do occur, they are more likely to form over the outlying foothills and higher terrain east of the Sound, where colder air aloft can interact more readily with surface heating.

Cold-core funnel clouds are occasionally observed in these setups, and waterspouts can also develop over Puget Sound during periods of spring instability. These events are usually brief and weak, but they serve as reminders that even here, the atmosphere can surprise us.

April usually greets us with gentle rain and fresh growth, but it also carries reminders of the atmosphere's complexity. Most years pass without incident, defined more by blossoms than by headlines. Still, April asks to be taken seriously, even as it invites optimism.

It's a month that rewards attention rather than certainty—offering progress without promises, warmth without guarantees, and just enough instability to keep us watching the sky.

April doesn't rush the season forward; it simply shows us that change is underway.

QUICK TAKEAWAY

April at a glance

- Showers replace storms

- Sunbreaks are longer and warmer

- Outdoor plans become more flexible

- Spring feels established, not tentative

May

The Promise Month

May feels like a deep breath. After months of watching the sky closely, this is when the Pacific Northwest begins to relax. The days are longer now, the light is warmer, and the weather more often cooperates than not. While nothing is guaranteed, May carries an unmistakable sense of possibility.

Rain hasn't disappeared, but it has changed character. Showers are lighter, more scattered, and easier to work around. Mornings often begin cool and gray, then give way to brighter afternoons. When the sun breaks through in May, it lingers. The landscape responds quickly, filling in with color and growth that feels almost sudden.

May is also the month when patience is rewarded. Outdoor plans become more reliable, windows stay open longer, and evenings stretch comfortably toward night. For many, this is when the Northwest feels most balanced—not too hot, not too cold, just right.

Seattle & Portland

May weather in Seattle and Portland often feels like a turning point. Seattle sees fewer prolonged rain events, replaced by a mix of sunbreaks and passing showers. When the clouds part, the combination of longer daylight and higher sun angle makes afternoons feel noticeably warmer than earlier in the spring.

Portland typically experiences slightly warmer conditions, with an increasing number of dry days as May progresses. East winds through the Columbia Gorge are rare now, and cold air intrusions are largely behind us. Rainfall still occurs, but it tends to be brief and less disruptive.

May is when the Pacific Northwest feels fully engaged with the outdoors again. Days are longer, rain is less persistent, and confidence returns to weekend plans. While showers still occur, they're usually brief enough to work around rather than cancel plans entirely.

In Seattle, waterfront activity increases noticeably. Walks along Elliott Bay, Alki Beach, and Lake Union become regular routines rather than exceptions. Boat traffic picks up, kayaks return to the water, and parks fill with people enjoying longer afternoons. Views of the Olympics and Cascades become more frequent as skies clear more often. May is also a popular time for visits to local gardens and arboretums, where spring color is at its peak without the crowds of summer.

In Portland, May marks the start of dependable outdoor living. Forest Park trails dry out and invite longer hikes, while neighborhood parks host picnics, casual sports, and unhurried walks. The Willamette River waterfront becomes a daily draw for runners, cyclists, and evening strollers. Farmers markets are fully active now, offering fresh produce and signaling the arrival of the growing season.

May is also when outdoor events feel comfortable again—community gatherings, youth sports, gardening projects, and simple backyard evenings all return to the rhythm of daily life. It's a month that encourages participation rather than observation.

May dressing in the Pacific Northwest shifts toward comfort and versatility rather than protection from the elements.

In both Seattle and Portland, light layers work best. Mornings can still feel cool, especially near the water, while afternoons often warm comfortably under partial sunshine. A lightweight jacket or sweater is usually enough, with a thin waterproof layer kept close at hand for passing showers.

Footwear can finally lighten up as well. Comfortable walking shoes are often sufficient, though water-resistant options remain helpful for damp trails or shaded paths. By May, sidewalks and park paths dry more quickly, making outdoor time feel easier and less planned.

Sunglasses become a regular companion, as sunbreaks are brighter and more frequent. Hats are useful during longer outings, especially as the higher sun angle brings warmth without warning.

Umbrella habits remain unchanged—common in Portland, less so in Seattle—but May rewards those who travel light and adapt as the day unfolds.

By the end of the month, jackets are often carried rather than worn, a quiet sign that summer is beginning to take shape.

If you like numbers, here's what May usually looks like.

Seattle
- Average high: 65.3 degrees F

- Average low: 48.3 degrees F

- Average mean temperature: 56.8 degrees F

- Average monthly rainfall: 2.16 inches

Extremes:

- Maximum: 92 degrees F, 1963

- Minimum: 35 degrees F, 2002

- Precipitation: 4.67 inches, 1948

Portland

- Average high: 69.3 degrees F

- Average low: 49.4 degrees F

- Average mean temperature: 59.4 degrees F

- Average monthly rainfall: 2.51 inches

Extremes:

- Maximum: 100 degrees F, 1983

- Minimum: 29 degrees F, 1954

- Precipitation: 5.55 inches, 1998

May doesn't eliminate uncertainty, but it tilts the odds in your favor.

WEATHER LORE

"A wet May makes a big load of hay."

This old saying reflects a simple truth: spring moisture matters. In the Pacific Northwest, May rain helps sustain growth before the long dry stretch of summer arrives. While many people hope for endless sunshine, a little rain now often pays dividends later.

May teaches moderation. Too much rain can feel inconvenient, but just enough helps keep forests green, rivers flowing, and gardens thriving well into the season.

"Cast not a clout till May be out. "

This old saying warns against putting winter clothes away too soon, and in the Pacific Northwest it still earns respect. May often delivers warm afternoons that invite lighter clothing, but cool evenings and surprise showers are never far behind. The saying isn't about clinging to winter—it's about pacing ourselves. May rewards optimism, but it also favors a sweater kept within reach.

"A wet May makes a full barn."

In an agricultural sense, May moisture is a gift. Rain that falls now feeds pastures, gardens, and forests just as growth accelerates. In the Northwest, a damp May often leads to greener hills, fuller rivers, and a more vibrant summer landscape. It may slow outdoor plans at times, but the payoff shows up later in ways that are easy to appreciate.

"If May be cold, the year will be bold. "

This saying hints that a cool start to May doesn't spell disappointment. Cooler springs often delay growth slightly, but they can also preserve soil moisture and moderate the heat of summer. In the Northwest, a restrained May often leads to a longer, more comfortable warm season. It's a reminder that early warmth isn't always the best measure of what lies ahead.

FROM THE WEATHER EYE

May is when the Northwest starts to feel generous. The air softens, the sky opens more often, and the outdoors becomes inviting again. Trails fill with hikers, patios reappear, and the rhythm of daily life shifts outward.

There's also a noticeable change in how people talk about the weather. Conversations move from complaints to observations. Rain becomes something to work around rather than endure. A cloudy morning no longer feels discouraging, because experience has taught us what the afternoon might bring.

I remember so many camping trips with the family over the Memorial Day weekend where it never stopped raining. If it did, I forgot. I guess rainy days in May bring memories.

May in Seattle often unfolds in stages. It's not unusual for the month to open on May 1st with a burst of warm, sunny days that feel more like early summer than spring.

Just as people begin to trust it, cooler and wetter weather frequently returns, reminding everyone that the season isn't finished settling in yet. Locals often call May 1st "the second spring"

This back-and-forth gives May the feeling of a second spring—one that arrives after the first false start. Temperatures swing, expectations reset, and patience is tested once more before steadier summer warmth finally takes hold later in June or July.

It's one of the reasons May keeps locals paying attention. The weather may look confident, but it's still deciding what it wants to be.

In Vancouver, Washington the annual Hazel Dell Parade of Bands occurs on the third Saturday of the month. Since the first parade began in 1964, rain has only fallen in a handful of years during parade time. It is one Saturday locals can rely on for fair weather.

On the warmest May day ever recorded in the Portland–Vancouver area, temperatures reached 100 degrees. The heat had an unexpected consequence when the Interstate 5 drawbridge opened for a routine boat passage. After the span was lowered, the steel expansion joints had expanded so much in the heat that the bridge could not close properly.

Traffic on both sides of the Columbia River quickly backed up for miles. With temperatures continuing to climb, a Portland fireboat was called into service and sprayed water onto the bridge deck to cool the steel. Once temperatures dropped enough for the metal to contract, the bridge was finally able to close and traffic resumed.

The date was May 28, 1983—a reminder that even in late spring, extreme heat can arrive suddenly and leave a lasting impression.

I think of May as the Northwest showing its best manners. The weather isn't perfect, but it's generous enough to remind us why we put up with the darker months. The promise feels real now—and most days, it's kept.

QUICK TAKEAWAY

May at a glance
- Longer, warmer days become the norm

- Showers are lighter and less frequent

- Outdoor plans grow more reliable

- Spring feels fully established

June

Waiting for Summer

June is the month that tests expectations. On the calendar, it belongs to summer. In reality, the Pacific Northwest often has other plans. The days are long now—remarkably so—but warmth can be inconsistent, and the sky frequently holds onto a stubborn layer of cloud.

This is the season of patience revisited. Mornings often begin gray and cool, especially near the coast and inland waters. Afternoons may brighten slowly, sometimes not until late in the day. When sunshine does arrive, it feels earned, and people make the most of it, knowing it may not last.

June isn't uncomfortable; it's cautious. Temperatures are mild, evenings cool, and the air carries a freshness that hasn't yet been baked away by summer heat. It's a month that asks people to slow down, adjust expectations, and let the season unfold in its own time.

Seattle & Portland

June weather highlights the subtle but important differences between Seattle and Portland. Seattle often remains under the influence of marine air, especially during the first half of the month. Morning cloud cover—sometimes lingering well into the afternoon—can keep temperatures cooler than expected. When skies do clear, the long daylight hours quickly warm sidewalks and storefronts.

Portland tends to see more frequent afternoon sunshine, though June clouds still make regular appearances. Temperatures rise more easily once skies break, but evenings cool off quickly. The influence of the Pacific remains strong, keeping extremes in check and preventing true summer heat from settling in just yet.

June is the month when outdoor plans finally feel dependable. Daylight stretches well into the evening, rain becomes less frequent, and the atmosphere feels cooperative rather than negotiable. It's a month that invites longer stays outside and fewer backup plans.

In Seattle, waterfront spaces come alive in June. Walks along Elliott Bay, Alki Beach, and Lake Union extend later into the evening, often under lingering twilight. Kayaking, paddleboarding, and small-boat activity increase as waters calm and temperatures moderate. Parks and viewpoints draw steady crowds, especially on clearer days when mountain views make brief but memorable appearances. June is also a favored time for outdoor concerts, neighborhood festivals, and evening gatherings that take advantage of the long light without summer's peak heat.

In Portland, June marks the start of true outdoor rhythm. Forest Park trails are at their best—green, shaded, and comfortably cool. The Willamette River waterfront becomes a nightly destination for walkers, runners, and cyclists enjoying extended daylight. Picnics, backyard gatherings, and community events fill weekends, while farmers markets reach peak variety. June evenings often linger outdoors well past dinner, a habit that will carry through the warmer months ahead.

Portland's Rose Festival typically gets underway during the first week of June, drawing thousands of visitors to daily events that culminate with the Grand Floral Parade. It's a celebration that signals the city's shift toward summer—most years.

Occasionally, however, an upper-level low pressure system slides south along the British Columbia coast and settles offshore near the southern Washington or northern Oregon coast. When this happens, bands of moisture rotate inland, bringing repeated rounds of showers—and sometimes steady rain—that can linger for hours at a time.

When that pattern sets up, the festival grounds along the Willamette River can quickly turn into a mud fest. Parade-goers improvise as best they can, seeking shelter wherever it's found, while umbrellas, ponchos, and good humor become essential accessories.

Locals have a name for this familiar interruption: the Rose Festival Low. It's well known, quietly dreaded, and notoriously difficult for forecasters to pin down. Some years it never appears. Other years, it makes its presence felt just in time for opening day.

It's a reminder that in early June, summer may be approaching—but the atmosphere still has a say in how celebrations unfold.

Across both cities, June encourages participation rather than observation. It's a time for lingering walks, unhurried conversations, and making plans that no longer hinge entirely on the forecast.

June dressing in the Pacific Northwest finally leans toward comfort, though a bit of caution still pays off.

In both Seattle and Portland, lighter clothing becomes the norm during the day, especially under extended sunshine. Short sleeves are common by afternoon, but mornings and evenings can still feel cool, particularly near the water. A light jacket or sweater remains useful for early starts and lingering twilight hours.

Footwear can be fully seasonal now. Comfortable walking shoes or light hikers work well for city paths, parks, and trails. As conditions dry out, waterproof shoes are less critical, though they're still helpful for shaded areas or early morning grass.

Sun protection becomes more important in June. Sunglasses are essential, and hats are increasingly common during longer outdoor outings. While rain is less frequent, a thin layer that handles an occasional shower is still worth keeping nearby.

By late June, the focus shifts from staying dry to staying comfortable. Layers are still part of the routine, but they're chosen for flexibility rather than necessity—a quiet sign that summer has arrived. Sunglasses tend to spend as much time waiting as they do being worn.

If you like numbers, here's what June usually looks like.

Seattle
- Average high: 70.2 degrees F

- Average low: 52.7 degrees F

- Average mean temperature: 61.4 degrees F

- Average monthly rainfall: 1.57 inches

Extremes:
- Maximum: 107 degrees F, 2021

- Minimum: 40 degrees F, 1924

- Precipitation: 3.70 inches, 1917

Portland
- Average high: 74.3 degrees F

- Average low: 54.1 degrees F

- Average mean temperature: 64.2 degrees F

- Average monthly rainfall: 1.63 inches

Extremes:
- Maximum: 116 degrees F, 2021

- Minimum: 39 degrees F, 1966

- Precipitation: 4.27 inches, 2010

June statistics suggest summer, but the sky often prefers to ease into it.

WEATHER LORE

"A cold June, a hot August."
Sayings like this capture the hope that early restraint will be rewarded later. In the Northwest, June coolness is often linked to marine influences rather than long-term pat-

terns, but the idea persists. When June holds back, people begin looking ahead, trusting that summer will eventually arrive in full.

Whether or not the saying proves true, it reflects a familiar mindset. June teaches patience, reminding us that warmth gained gradually often lasts longer.

"A calm June puts the farmer in tune."

June often brings a stretch of settled weather, and when it does, it's a welcome shift after spring's uncertainty. Calmer days allow planting, growing, and outdoor routines to fall into rhythm. In the Northwest, a quiet June usually signals that the atmosphere is transitioning toward summer patterns, even if clouds still linger along the coast or overnight.

"If June be fair, harvest comes early."

This saying reflects the idea that steady warmth and moderate moisture in June support healthy growth. In the Northwest, a fair June doesn't mean hot or dry—it means balanced. When rain tapers gradually and temperatures remain mild, forests deepen in green and gardens respond quickly. It's the kind of month that sets the tone for what follows.

"June fog brings fair weather."

Morning clouds and low fog are common in early summer, especially near water. While they may seem gloomy at first, they often signal stable conditions offshore. In many cases, the fog lifts by midday, giving way to brighter skies and comfortable temperatures. It's one of June's quieter patterns, and one that locals learn to trust.

FROM THE WEATHER EYE

June is often the transition month between spring's variability and summer's routines. While rainfall decreases overall, it rarely disappears all at once. Instead, showers become more scattered and lighter, especially west of the Cascades. Many Junes feel cooperative rather than settled—pleasant enough to plan around, but still capable of surprise.

Early summer cloudiness is another quiet June feature. Marine air from the Pacific often pushes inland overnight, bringing low clouds or morning fog to coastal areas and inland valleys. These clouds can linger into midday before gradually thinning, especially near Puget Sound and the lower Columbia. Locals learn not to judge the day too early; June mornings often improve with patience.

June also introduces longer dry spells, particularly late in the month. These stretches allow soils to dry, rivers to settle, and outdoor routines to take hold. It's when the

Northwest begins easing into its summer rhythm, even if temperatures remain moderate.

Wind patterns shift as well. The stronger frontal winds of spring fade, replaced by afternoon breezes driven more by heating differences than storms. Along the coast and through the Columbia Gorge, winds can still be noticeable, especially during warm inland days.

Finally, June carries an important balance point. A cooler, cloudier June often preserves moisture and leads to a more comfortable summer. A hotter, drier June can accelerate snowmelt, dry fuels early, and change expectations for what lies ahead.

June feels like the Northwest clearing its throat before speaking. The landscape is ready. Gardens are thriving, rivers run steady, and daylight stretches late into the evening. The only thing missing is consistency.

This is when conversations about the weather shift again. People stop asking if spring will arrive and start wondering when summer will finally show up. A bright afternoon feels encouraging, while a cool, gray morning can feel like a setback—even though both are perfectly normal.

There's also comfort in June's restraint. The air remains fresh, nights are ideal for sleeping, and outdoor activities feel easier without heat pressing down. Trails are inviting, windows stay open longer, and evenings seem designed for lingering.

Did you know that the month of June has more cloudy days and less sunshine than May? That is due to the ever-increasing heat over the inland Northwest. This draws in marine air and low cloudiness from the cooler Pacific waters. Typical June forecast for both Seattle and Portland: Morning clouds with afternoon sunshine.

I think of June as the month that prepares us for what's coming. It holds back just enough to make summer feel like a reward rather than an assumption. When warmth does arrive consistently, we're ready for it.

QUICK TAKEAWAY

June at a glance

- Long days, cooler mornings

- Marine clouds often delay sunshine

- Summer warmth arrives gradually

- Patience pays off later

July

DRY CONFIDENCE

July arrives with confidence. By now, the Pacific Northwest has usually made up its mind. Skies turn reliably blue, rain becomes a memory rather than a forecast, and the rhythm of daily life shifts fully outdoors. July doesn't hesitate—it settles in.

This is the heart of the dry season west of the Cascades. Storm systems retreat northward, cloud cover thins, and sunshine becomes the rule rather than the exception. Mornings often begin cool and comfortable, followed by warm afternoons that feel earned rather than oppressive. Evenings linger long past dinner, inviting walks, gatherings, and unhurried conversations.

July weather is dependable in a way that feels reassuring. After months of watching the sky closely, people relax their vigilance. The need for layers fades, windows stay open overnight, and plans are made with confidence rather than contingency.

Seattle & Portland

July weather in Seattle and Portland is remarkably similar, though small differences remain. Seattle often enjoys cooler afternoons, influenced by nearby water and occasional marine breezes. Morning clouds are still possible early in the month, but they tend to burn off quickly, leaving behind clear skies and mild temperatures.

Portland typically runs warmer, especially during the afternoon hours. The absence of cloud cover allows temperatures to climb more easily, though evenings usually cool off nicely. Heat waves do occur, but they are usually brief and followed by relief rather than persistence.

July is when the Pacific Northwest settles into its most dependable outdoor rhythm. The days are long, evenings stretch well past dinner, and weather plans rarely need a second thought. It's the month when people stop watching the forecast so closely and simply step outside.

In Seattle, July invites time near the water. Lake Washington and Lake Union draw swimmers, paddleboarders, and small sailboats enjoying warm afternoons and light breezes. Evening walks along neighborhood viewpoints become routine as sunsets linger, and outdoor movies, music, and informal gatherings fill parks and open spaces. Clearer skies often reward patient observers with long-distance mountain views that seem sharper and more frequent this time of year.

In Portland, July encourages early starts and late finishes. Mornings are popular for walks and runs before the day warms, while evenings bring people back outdoors for neighborhood strolls, outdoor dining, and riverfront activity. The Willamette and Columbia rivers become a focal point for floating, kayaking, sailing and relaxed shoreline visits. Local festivals and community events are common, taking full advantage of predictable weather and extended daylight.

Across both cities, July favors unstructured outdoor time—lingering conversations, backyard evenings, and simple routines that stretch longer than planned. It's a month less about seeking novelty and more about enjoying consistency, a rare and welcome feature in Northwest weather.

In both Seattle and Portland, light, breathable clothing is the norm during the day. Short sleeves and lightweight fabrics work well, especially during afternoon warmth. Even so, temperatures rarely feel oppressive, particularly near water or in shaded areas.

Evenings can still cool off, especially after sunset. A light layer—a thin jacket or long-sleeve shirt—is often enough to stay comfortable during waterfront walks or backyard gatherings.

Footwear is fully summer-ready. Comfortable walking shoes, sandals, or light trail shoes suit most July activities, whether in the city or along rivers and paths.

Sun protection becomes more important this month. Sunglasses are a daily companion, and hats are common during longer outings. While rain gear usually stays home, it's not unusual for locals to keep a light layer nearby more out of habit than necessity.

Umbrellas disappear entirely, replaced by hats and water bottles.

July dressing is about ease. Once outside, there's rarely a need to adjust plans—or clothing—for the weather.

If you like numbers, here's what July usually looks like.

Seattle
- Average high: 76.5 degrees F

- Average low: 56.5 degrees F

- Average mean temperature: 66.5 degrees F

- Average monthly rainfall: .78 of an inch

Extremes:

- Maximum: 105 degrees F, 2009

- Minimum: 44 degrees F, 1902

- Precipitation: 2.36 inches, 1897

Portland

- Average high: 81.9 degrees F

- Average low: 58.5 degrees F

- Average mean temperature: 70.2 degrees F

- Average monthly rainfall: .51 of an inch

Extremes:

- Maximum: 107 degrees F, 1965

- Minimum: 43 degrees F, 1955

- Precipitation: 2.68 inches, 1983

July's statistics tell a simple story: dry, warm, and reliable.

WEATHER LORE

"When July is dry, August will reply."

This saying hints at the balance people expect from summer weather. In the Northwest, July dryness is normal rather than predictive, but the sentiment endures. A dry July sets the tone for the season, establishing patterns that often carry through much of the summer. Whether August responds or not, July usually delivers what it promises.

"Thunder in July brings warm weather by."

In much of the country, July thunderstorms are common. In the Pacific Northwest, they're the exception. When thunder does occur in July, it often signals a brief intrusion of unstable air into an otherwise stable summer pattern. More often than not, it's followed by warmer, calmer conditions rather than prolonged unsettled weather. It's one of those moments when the atmosphere briefly reminds us it hasn't gone entirely quiet.

"If July be dry, winter will be hard."

This saying is often debated, but it reflects a long-standing concern about balance. In the Northwest, a very dry July can dry soils early, stress vegetation, and raise concerns about fire season. While a dry July doesn't predict winter severity, it does shape how the rest of the warm season unfolds. Locals pay attention—not out of worry, but awareness.

"July suns are strong, though the air feels mild."

July sunshine in the Northwest can be deceptive. Temperatures often remain comfortable, especially near water, but the higher sun angle delivers strong midday rays. It's why sunburns can arrive faster than expected, even on days that don't feel hot. July teaches respect for the sun without demanding retreat from it.

FROM THE WEATHER EYE

July feels like a reward. After months of watching for sunbreaks and counting dry hours, the weather finally cooperates. The air is clear, visibility stretches for miles, and familiar landmarks—mountains, rivers, and coastlines—stand out sharply against the sky.

There's a noticeable change in how people move through the day. Schedules loosen. Evenings stretch later. Outdoor events fill the calendar, and weather becomes something to enjoy rather than manage.

At the same time, July asks for a different kind of awareness. Dry conditions mean rivers drop, soils harden, and landscapes grow increasingly vulnerable. Wildfire risk becomes part of the conversation, especially east of the Cascades and during extended heat.

Still, July remains generous. The balance of warmth and comfort, light and cool nights, makes it one of the most forgiving months of the year. Confidence replaces caution, and the Northwest settles into its summer stride.

Most native residents use the phrase, "Summer begins on July 5". Rightly so. Statistics show warmer temperatures are likely after the Fourth of July. I have spent many years outside after sunset watching the fireworks wrapped in a warm blanket Besides keeping warm, it helps with the pesky mosquitoes.

July is when the Northwest weather settles into a steady rhythm. Storm systems retreat north, the jet stream weakens, and the atmosphere finally seems content to leave us alone. Days often begin clear or lightly hazy and end the same way, with only minor variations from one to the next.

That consistency is what makes July feel different. After months of watching the sky closely, people relax their attention. Forecasts matter less. Plans stick. It's the one month when the weather rarely asks for compromise.

And yet, July isn't entirely without personality. Heat waves can arrive suddenly, especially east of the Cascades, and occasionally push west into Portland or Seattle. These events usually last only a few days, but they remind us that even a mild climate has its limits. When temperatures climb, nights remain warmer than usual, and relief comes slowly.

More often, July offers balance. Warm afternoons give way to cooler evenings, especially near water. Breezes replace storms, and long daylight does the rest. It's a month that rewards simplicity—and asks little in return.

Portland's hottest temperatures have often occurred in July, even though August is usually thought of as the peak of summer. The city's all-time record highs have tended to cluster during short-lived July heat waves rather than extended late-summer heat.

Seattle's July rainfall totals are typically among the lowest of the year, but measurable rain still happens. A light shower or brief marine push can interrupt an otherwise dry stretch, usually arriving overnight or early in the morning and disappearing quickly.

In July, Northwest residents stop asking what the weather will do and start assuming it will behave. Most of the time, they're right.

QUICK TAKEAWAY

July at a glance
- Dry weather dominates

- Warm afternoons, cool nights

- Outdoor plans are dependable

- Summer feels established and confident

August

Blue Skies and Smoke

August often begins with confidence carried over from July. Skies remain blue, rain is scarce, and summer feels firmly in control. Days are warm, sometimes hot, and evenings still linger long enough to invite outdoor plans that stretch late into the night.

This is the driest month of the year across much of the Pacific Northwest. Storm systems remain far to the north, and the atmosphere settles into a predictable pattern. Mornings start clear, afternoons warm steadily, and nights cool just enough to bring relief. It's a rhythm that feels dependable—until it doesn't.

August also carries a growing awareness. As the month progresses, landscapes dry further, rivers drop, and forests become increasingly vulnerable. The same sunshine that defines August can, under the wrong conditions, introduce smoke and haze that soften the sky and mute the distance.

Seattle & Portland

August weather in Seattle and Portland is often remarkably similar. Both cities experience extended dry periods, clear skies, and warm afternoons. Seattle typically runs slightly cooler, moderated by nearby water and occasional onshore breezes that temper the heat.

Portland generally sees warmer daytime temperatures, especially during heat waves. Nights, however, still cool efficiently, offering some relief after hot afternoons. Rain remains unlikely in both cities, and when it does appear, it's usually brief and light.

August is the height of summer in the Pacific Northwest. Rain is scarce, daylight remains generous, and outdoor life shifts toward water, shade, and evening hours. The pace slows slightly, not from weather uncertainty, but from warmth and habit.

In Seattle, August revolves around water and long evenings. Lake Washington, Lake Union, and Puget Sound draw swimmers, boaters, and paddleboarders seeking relief from warm afternoons. Beaches and shoreline parks remain busy well into the evening as temperatures ease and sunsets stretch late. Outdoor festivals and neighborhood events are common, with Seafair festivities standing out as a signature late-summer tradition that brings crowds to the waterfront and skies.

In Portland, August outdoor life often shifts to early mornings and after-dinner hours. River access becomes a focal point, with floating, kayaking, and quiet shoreline gatherings along the Willamette. Shaded trails and forested parks are favored during the warmest part of the day, while evenings bring people back outdoors for concerts, movies, and community events. Late-summer festivals and outdoor markets continue, often paired with warm, dry weather that feels dependable.

Across both cities, August encourages unstructured outdoor time. Plans are simple, routines familiar, and the weather usually stays out of the way. It's a month defined less by novelty and more by ease.

August dressing in the Pacific Northwest is straightforward and summer-focused, with attention shifting from layers to comfort and sun protection.

In both Seattle and Portland, lightweight clothing is standard. Short sleeves, breathable fabrics, and loose fits help manage warm afternoons, especially away from water or shade. Footwear is fully seasonal, with sandals, light shoes, and casual walking shoes covering most activities.

Evenings remain comfortable but can cool slightly, particularly near rivers or along the Sound. A very light layer may be useful after sunset, though many nights pass without it.

Sun protection is important in August. Sunglasses and hats are common, and sunscreen becomes part of daily routine during longer outings. Rain gear is rarely needed and often left behind altogether.

August dressing reflects the month itself—simple, relaxed, and predictable. Once outside, there's little need to adjust for changing conditions, a quiet luxury in Northwest weather.

If you like numbers, here's what August usually looks like.

Seattle
- Average high: 77 degrees F

- Average low: 57.1 degrees F

- Average mean temperature: 67.1 degrees F

- Average monthly rainfall: 1 inch

Extremes:
- Maximum: 97 degrees F, 2020

- Minimum: 46 degrees F, 1908

- Precipitation: 5.49 inches, 1977

Portland

- Average high: 82.3 degrees F

- Average low: 58.9 degrees F

- Average mean temperature: 70.6 degrees F

- Average monthly rainfall: .54 of an inch

Extremes:

- Maximum: 108 degrees F, 2023

- Minimum: 44 degrees F, 1980

- Precipitation: 4.53 inches, 1968

August statistics highlight summer's peak—but they don't tell the whole story.

WEATHER LORE

"If August be fair, winter will be harsh."

Sayings like this reflect an old habit of looking ahead too far. In reality, August weather has little to say about winter's temperament. Still, the saying captures something familiar: when August is especially beautiful, people begin to wonder what balance might follow.

In the Northwest, August reminds us to enjoy what's in front of us without borrowing concern from seasons yet to come.

"Dry August and warm, doth harvest no harm."

This saying reflects the value of a settled August. In the Pacific Northwest, dry weather during this month allows crops to mature, hay to cure, and outdoor routines to proceed without interruption. While dryness brings its own concerns, a calm August has long been seen as a necessary bridge between summer growth and autumn change.

"If August be clear, expect a fertile year."

Clear skies in August often signal a stable atmosphere and a productive season nearing completion. In the Northwest, sunshine this time of year deepens late-summer warmth without the volatility of earlier months. Forests dry, rivers settle, and the landscape pauses before the coming shift. It's a moment of balance rather than excess.

"In August, the sun lingers, and so do people."

This isn't an old proverb so much as an observation that feels timeless. August invites lingering—on porches, along riverbanks, and in the quiet moments after sunset when the day finally cools. In the Northwest, this is when summer feels most settled and least hurried. Evenings stretch, conversations drift, and there's little urgency to head indoors. August doesn't rush us along; it encourages us to stay awhile.

"August sun is strong, though the days grow shorter."
Even as daylight begins to fade, August sunshine remains powerful. In the Northwest, temperatures may feel moderate, especially near water, but the sun's intensity can be misleading. This old saying reminds us that summer's strength lingers, even as subtle signs of change begin to appear.

FROM THE WEATHER EYE

August often feels suspended in time. The weather is steady enough that days blur together, and the absence of rain becomes unremarkable. Lawns fade, dust rises on dry roads, and the landscape settles into muted tones. It's not drought yet—just a pause.

Haze becomes part of the August sky, even when no smoke is nearby. Warm air, long days, and distant influences soften the horizon, making mountains appear less crisp than earlier in summer. Locals notice the change instinctively, reading the sky not for storms but for clarity.

I will say that in my recent observations over the past decade the Pacific Northwest has had regular wildfire smoke intrusion. Some from Siberia. Others from Canada. Offshore winds usher in smoke from wildfires in eastern Washington and Oregon. And let's not forget smoke from our neighbors in California. Each year August and into September bring wildfire smoke. Usually it is high aloft but occasionally it mixes to the lower elevations where we breathe. Air quality inches up into the moderate and unhealthy ranges.

Nights begin to tell the story of the season turning. Even during warm afternoons, overnight cooling becomes more reliable. Windows stay open longer, fans cycle less, and early mornings feel calmer. It's one of the first easy to miss signs that summer is no longer accelerating.

August is also when people adjust their routines without thinking about it. Outdoor activity shifts earlier in the day or later in the evening. Midday becomes quieter, not because plans are canceled, but because they're rescheduled. The weather hasn't forced the change—it's simply suggested it.

And despite its reputation for sameness, August can still surprise. A brief marine push can cool coastal areas overnight. A short heat wave can raise inland temperatures unexpectedly. These moments stand out precisely because the rest of the month feels so consistent.

August feels expansive. Visibility stretches for miles on clear days, and familiar landmarks stand out sharply against the sky. Mountains, rivers, and coastlines appear closer than they are, framed by long hours of steady light.

At the same time, August asks for awareness. Dry conditions mean landscapes are fragile, and smoke from regional wildfires can drift in unexpectedly. Skies that were blue the day before may turn hazy overnight, softening the sun and muting colors. These episodes can feel unsettling, but they're now part of late-summer reality.

When the air is clear, August is generous. Trails are dry, water levels stable enough for recreation, and evenings are ideal for lingering outdoors. The month invites enjoyment—but with an understanding that summer's peak also carries responsibility.

I think of August as the season at full stride, aware of its own power. It's a month to appreciate clear skies, respect dry ground, and recognize that change is already waiting just beyond the horizon.

QUICK TAKEAWAY

August at a glance
- Dry weather dominates

- Warm days, cooler nights

- Smoke may affect visibility

- Summer reaches its peak

September

Gentle Letting Go

September arrives softly. Summer doesn't end all at once in the Pacific Northwest—it eases away. The light shifts first. Afternoons shorten, shadows lengthen, and evenings cool just enough to suggest a change without insisting on it.

This is often one of the most comfortable months of the year. Warm days still appear, sometimes generously so, but the heat loses its edge. Nights grow cooler and more restful, windows stay open, and mornings carry a hint of crispness that feels refreshing rather than cold.

September is also a month of balance. The dry spell of summer often continues, but the atmosphere feels less rigid. Skies remain clear more often than not, yet the return of clouds no longer feels unwelcome. The season begins to breathe again.

Seattle & Portland

September weather in Seattle and Portland tends to reward patience. Seattle often enjoys clear days with cooler afternoons, especially as marine air becomes more influential later in the month. Morning clouds may return at times, but they usually lift, leaving behind comfortable temperatures and clean air.

Portland typically sees warmer afternoons early in September, followed by a gradual cooling trend as the month progresses. Heat waves become less likely, and evenings cool quickly once the sun sets. The long summer dryness often holds, but the first light rain event is always possible, sometimes arriving unexpectedly after weeks of blue skies.

September is a favorite month for outdoor time in the Pacific Northwest. Crowds thin, temperatures moderate, and the pace of life slows just enough to notice.

In Seattle, waterfront walks become especially appealing as the air cools and skies often remain clear. Views sharpen, mountain outlines return, and evening light takes on a softer quality. Parks, neighborhood paths, and shoreline benches invite longer stays without summer's bustle. September is also a popular time for casual boating and late-season paddle outings, when conditions are calm and predictable.

In Portland, September encourages unhurried exploration. Forest Park trails feel cooler and more comfortable, and the Willamette River waterfront draws walkers and cyclists enjoying milder afternoons. Farmers markets shift toward harvest offerings, and outdoor dining continues comfortably well into the evening.

Across both cities, September favors reflection over activity for activity's sake. It's a month for lingering walks, quieter gatherings, and noticing the subtle turn toward autumn without feeling rushed.

September dressing in the Pacific Northwest shifts back toward layers, though summer hasn't fully let go.

In both Seattle and Portland, afternoons often remain warm enough for short sleeves, especially in the sun. Mornings and evenings, however, cool quickly. A light jacket or sweater becomes part of the routine again, particularly near water or after sunset.

Comfortable walking shoes remain the standard, though shaded trails and early-morning grass may feel damp. Sunglasses are still useful, but hats and sunscreen begin to appear less frequently as the sun's intensity softens.

Rain gear slowly returns to closets and car trunks. While September often stays dry for long stretches, the first passing showers can arrive with little notice. Dressing for September means preparing for variety rather than committing to a season.

If you like numbers, here's what September usually looks like.

Seattle
- Average high: 71.3 degrees F

- Average low: 53.2 degrees F

- Average mean temperature: 62.3 degrees F

- Average monthly rainfall: 1.74 inches

Extremes:
- Maximum: 92 degrees F, 2025

- Minimum: 36 degrees F, 1908

- Precipitation: 5.62 inches, 1978

Portland
- Average high: 76.7 degrees F

- Average low: 54.1 degrees F

- Average mean temperature: 65.4 degrees

- Average monthly rainfall: 1.52 inches

Extremes:

- Maximum: 105 degrees F, 1988

- Minimum: 34 degrees F, 1965

- Precipitation: 5.62 inches, 2013

September doesn't rush change, it introduces it.

WEATHER LORE

"September sun warms the heart, October frost the ground."
This saying captures September's gift perfectly. The warmth that remains feels gentler, more personal, as if offered rather than imposed. While frost still lies ahead, September allows time to enjoy what's left before cooler days arrive in early October.

"September blow soft, till the fruit's in the loft."
This old saying values gentler weather as harvest season approaches. In the Pacific Northwest, September often delivers just that—calm days, cooler nights, and fewer extremes. It's a month that allows gardens, orchards, and vineyards to finish strong without the stress of heat or storms.

"September sun is gentle, but its warmth is real."

The sun sits lower in the sky now, and its strength is easier to tolerate. Afternoons still feel warm, but without summer's intensity. In the Northwest, this balance makes September one of the most comfortable months of the year, especially for time spent outdoors.

"As September goes, so goes the fall."

This saying reflects the idea that September sets the tone for the season ahead. A dry, mild month often eases the transition into autumn, while an early return of rain signals that the long dry stretch is ending. Either way, September carries the first real hints of change.

FROM THE WEATHER EYE

September often feels like the Northwest is at ease with itself. The rush of summer activity slows, schedules loosen, and the outdoors become less crowded. Trails feel quieter, shorelines calmer, and evenings more inviting for reflection.

Weather-wise, September brings clarity. Skies often appear sharper, visibility improves, and distant landmarks stand out again after summer haze. When clouds return, they tend to arrive gently, reminding us that change is underway without demanding immediate adjustment.

There's also a gradual emotional shift. September invites pause. It encourages appreciation rather than anticipation. Summer has delivered what it promised, and now the season begins to close its chapter calmly.

I think of September as the month that teaches us how to let go gracefully. It doesn't ask us to abandon summer, only to notice that the light is changing and that something quieter—and equally meaningful—is approaching.

September nights are different — not cold, just decided.

Even after warm afternoons, nighttime cooling becomes more reliable. Windows stay open longer, but blankets reappear. It's the first time since spring that evening air feels intentional rather than incidental. The day may still belong to summer, but the night begins to claim autumn.

September often marks the emotional end of the dry season, even if rain hasn't returned yet.

Lawns are tired, dust coats side streets, and the smell of dry leaves replaces cut grass. People begin listening for rain again, not out of urgency, but expectation. September doesn't always deliver it — but it reminds us that the dry stretch has a limit.

September doesn't announce change; it lets us notice it.

QUICK TAKEAWAY

September at a glance
- Comfortable days, cooler nights

- Dry weather often lingers

- First hints of fall appear

- Change arrives gently

October

Turning Inward

October marks a clear change. The light lowers, the days shorten noticeably, and the Pacific Northwest begins to turn inward again. Summer's confidence has faded, replaced by a quieter, more deliberate rhythm. The outdoors remains inviting, but it no longer insists.

This is the month when autumn truly arrives west of the Cascades. Rain returns—not all at once, but with growing intention. Early October often begins gently, with scattered showers and breaks of calm weather. As the month progresses, rain becomes more frequent, nights grow cooler, and mornings take longer to brighten.

October doesn't rush. It eases the region back into its familiar fall patterns, reminding residents that the dry season was always temporary. The return of rain feels expected, even welcome, after months of restraint.

Seattle & Portland

October weather in Seattle and Portland reflects the season's transition. Seattle often sees an increase in steady, light rain events, with clouds becoming more persistent. Sunbreaks still occur, especially early in the month, but they feel less reliable. Marine air plays a stronger role now, keeping temperatures mild but damp.

Portland experiences a similar return to wetter conditions, though early October can still offer warm, dry afternoons before the shift fully settles in. East winds through the Columbia Gorge are uncommon, and temperatures remain moderate. As rain returns, the city settles comfortably back into its fall routine.

October is when the Pacific Northwest turns inward again, but not indoors just yet. Cooler air, returning rain, and shorter days shift outdoor time toward quieter, more deliberate outings.

In Seattle, October favors neighborhood walks, waterfront paths, and city parks framed by autumn color. After rainfall, the air often feels especially clear, and views across Puget Sound sharpen between showers. Parks and greenways remain active, though crowds thin as evenings arrive earlier. October is also a popular month for shoreline walks and casual photography, as changing leaves and reflective surfaces transform familiar scenes.

In Portland, October brings a return to forested trails and sheltered parks. Forest Park becomes especially inviting as fall color spreads beneath the canopy and cooler temperatures make longer walks comfortable again. The Willamette River waterfront remains active on dry days, while farmers markets shift fully into harvest mode. Outdoor activity continues, but plans are now shaped around weather windows rather than assumed sunshine.

Across both cities, October encourages attentiveness. Outdoor time becomes more observational than social, shaped by light, moisture, and the rhythm of passing systems.

October dressing in the Pacific Northwest returns to layers with purpose.

In both Seattle and Portland, a reliable waterproof jacket becomes essential again. Showers are more frequent, and rain tends to linger longer than in early fall. Underneath, light sweaters or fleece layers provide warmth without bulk, especially during cooler mornings and evenings.

Footwear should handle wet pavement and muddy paths. Comfortable walking shoes with good traction are helpful, particularly in parks and forested areas. Umbrellas reappear more often in Portland, while Seattle residents continue to favor hoods and timing outings between showers.

Hats and gloves are not yet routine, but they begin to appear on cooler mornings. October clothing is about readiness rather than insulation—preparing for changing conditions without surrendering to winter.

If you like numbers, here's what October usually looks like.

Seattle
- Average high: 60.5 degrees F

- Average low: 46.7 degrees F

- Average mean temperature: 53.6 degrees F

- Average monthly rainfall: 3.65 inches

Extremes:
- Maximum: 88 degrees F, 2022

- Minimum: 29 degrees F, 1929

- Precipitation: 10.30 inches, 2016

Portland
- Average high: 64.4 degrees F

- Average low: 46.7 degrees F

- Average mean temperature: 55.6 degrees F

- Average monthly rainfall: 3.42 inches

Extremes:
- Maximum: 92 degrees F, 1987

- Minimum: 26 degrees F, 1921

- Precipitation: 8.41 inches, 1994

October statistics confirm what the sky already suggests—change is underway.

WEATHER LORE

"October winds bring winter in."
This saying reflects the feeling that October carries more weight than earlier fall months. Stronger systems begin to arrive, winds increase, and the atmosphere grows more active. While winter may still be weeks away, October signals that the transition has begun in earnest.

"When October brings the first frost, expect a wet winter."
This saying reflects the long-standing belief that an early chill signals a shift toward a more active storm season. In the Northwest, October frost does arrive occasionally,

especially away from water and higher terrain. While it doesn't guarantee what's ahead, it often marks the true end of the dry season.

"October wind lays the leaves at rest."

Autumn winds help clear trees and sidewalks alike. In October, passing fronts often arrive with gusty winds that strip remaining leaves and signal seasonal change. It's a visible reminder that summer's hold has loosened.

"A fair October gives a long winter's comfort."

This saying values balance. A calm, wet-but-not-stormy October allows soils to recharge and rivers to rise gradually. In the Northwest, a steady October often feels like the atmosphere easing into winter rather than rushing toward it.

FROM THE WEATHER EYE

October feels familiar. After months of dry skies and long evenings, the return of rain carries a sense of recognition. The sound of water on rooftops, the darkening of afternoons, and the return of clouds feel like old companions rather than intrusions. It is also the time when nature shows it glorious beauty as deciduous trees transition to their autumn colors. Vibrant yellows, red and orange splatter across the landscape. Fall storms whittle away until most foliage is gone. Some however, hang on into November.

This is also when people adjust their routines. Walks shorten, indoor spaces become more appealing, and the pace of daily life slows slightly. Outdoor plans continue, but they're made with flexibility again—an understanding that weather now has more say.

October rain often feels productive. Rivers respond quickly, soils soften, and landscapes regain their deep green. The air smells fresher, visibility improves between systems, and familiar landmarks take on sharper definition under cooler skies.

I think of October as the month when the Northwest settles back into itself. The season doesn't demand attention; it invites reflection. The shift inward feels natural, even comforting—a reminder that change doesn't always arrive abruptly, but often with quiet persistence.

October can still surprise. A late warm day may appear unexpectedly, or a stronger-than-expected storm may announce itself with gusty winds and heavy rain. But these moments feel appropriate now. The season welcomes them.

Vigorous cold fronts with unstable air aloft can produce those cold core funnels like we can see six months earlier in April. Some may briefly lower and touch the earth resulting in a tornado. Most are brief, most are weak. The coastal regions are most subjected to these and locals often spot waterspouts out into Pacific from shore.

Perhaps the most famous weather event this month is the Columbus Day windstorm of 1962. There had not been a windstorm that strong before and hasn't occurred since. Winds in Portland gusted to 116 mph on the Morrison Bridge. In Seattle, winds were clocked at 83 mph before power went out. Mt. Hebo in Oregon registered a gust to 176 mph. Renton had 100 mph, Bellingham 98 mph and Vancouver, Washington 92 mph. It killed 46 persons and knocked down enough timber to build one million homes. Power was out for a week or more in many places.

By month's end, the landscape settles into a new rhythm. Lawns green up, rivers respond, and the air carries a faint coolness that lingers even after rain has passed. October doesn't rush us indoors, but it does ask for adjustment.

It's a month of transition done quietly and well—one that prepares both land and people for what comes next.

October doesn't ask for attention. It simply earns it.

QUICK TAKEAWAY

October at a glance

- Rain returns gradually, then more steadily

- Days shorten noticeably

- Fall routines settle in

- The Northwest turns inward

November

The Long Middle

November settles in without much ceremony. The clocks have been turned back, daylight retreats quickly, and the Pacific Northwest returns to a familiar rhythm of clouds and rain. Autumn's color fades, replaced by bare branches and damp streets. November doesn't ask for attention—it simply arrives and stays.

This is often the wettest and most consistent month of the year west of the Cascades. Storms roll in from the Pacific with regularity, bringing periods of steady rain rather than dramatic bursts. The atmosphere feels occupied now, as if winter has taken over operations but hasn't yet revealed its full intent.

November weather is not severe, but it is persistent. Rain falls, pauses briefly, then resumes. Skies remain gray for days at a time, and temperatures hover comfortably cool rather than cold. It's a month that rewards acceptance more than resistance.

Seattle & Portland

November weather in Seattle and Portland is marked by consistency rather than surprise. Seattle often experiences extended periods of cloud cover and light to moderate rain, with only occasional breaks. Marine air keeps temperatures relatively mild, even as daylight diminishes.

Portland shares much of that pattern, though rainfall totals are often slightly higher. The return of frequent rain restores familiar sounds—the steady tapping on rooftops, water moving through gutters, leaves collecting along curbs. Cold air east of the Cascades remains mostly locked in place, and sharp temperature drops are uncommon.

In Seattle, waterfront walks remain popular, especially on calmer days between storms. Puget Sound often looks dramatic in November, with low clouds, shifting light, and passing showers adding depth rather than gloom. Neighborhood walks, urban parks, and short shoreline outings fit the season well, especially when paired with warm stops along the way.

In Portland, November favors forested parks and familiar routes. Forest Park remains active, though paths are quieter and more reflective. The Willamette River waterfront continues to draw walkers and runners during dry windows, while late-season farmers markets and neighborhood errands keep people moving outdoors despite the rain.

Across both cities, November encourages shorter, more intentional time outside. Walks are purposeful, pauses are appreciated, and the atmosphere itself becomes part of the experience. It's a month that doesn't demand much—just awareness and a good jacket.

November dressing in the Pacific Northwest is about staying dry and comfortable rather than warm alone.

In both Seattle and Portland, a dependable waterproof jacket becomes essential. Rain is more frequent and often longer-lasting, and dry breaks can be brief. Layering underneath allows adjustment as temperatures shift throughout the day.

Footwear matters more now. Shoes with good traction and water resistance are helpful on wet sidewalks, leaf-covered paths, and muddy trails. In Portland, umbrellas are commonly used and practical. In Seattle, many people continue to rely on hoods and timing their outings between showers.

Cool mornings and evenings make hats and light gloves more common, especially during longer walks. November clothing favors readiness and routine—items chosen not for fashion, but for reliability.

If you like numbers, here's what November usually looks like.

Seattle
- Average high: 51.9 degrees F

- Average low: 40.9 degrees F

- Average mean temperature: 46.4 degrees F

- Average monthly rainfall: 5.85 inches

Extremes:
- Maximum: 76 degrees F, 2010

- Minimum: 13 degrees F, 1985

- Precipitation: 11.56 inches, 2006

Portland
- Average high: 53.5 degrees F

- Average low: 40.6 degrees F

- Average mean temperature: 47.1 degrees F

- Average monthly rainfall: 5.45 inches

Extremes:

- Maximum: 73 degrees F, 1975

- Minimum: 13 degrees F, 1985

- Precipitation: 11.92 inches, 2006

November's statistics reflect endurance more than drama.

WEATHER LORE

"November rain is winter's refrain."

This saying captures November's repetitive nature. The rain returns again and again, not as a warning but as a reminder. Winter isn't here yet, but its voice is unmistakable.

"November rain makes fertile ground."

This saying reflects the practical side of late-fall weather. In the Pacific Northwest, November rain replenishes soils, fills streams, and restores balance after the dry months. While it can feel persistent, this moisture lays the groundwork for winter and the growing season that follows.

"When November fog sits low, winter is near."

Fog becomes more noticeable this time of year, especially during calm stretches between storms. Low clouds and morning fog often signal cooling nights and stable air. It's one of the quieter signs that autumn is giving way to winter.

"A warm November foretells a long winter."

This bit of lore hints at balance rather than prediction. In the Northwest, a mild November can delay the onset of colder weather, but it rarely cancels it. Winter still arrives—sometimes later, sometimes all at once. November reminds us that seasons don't always follow the calendar neatly.

FROM THE WEATHER EYE

November feels like the Northwest is settling into itself. The urgency of fall has passed, and winter hasn't yet demanded full attention. What remains is a stretch of time that asks for steadiness.

This is when people adapt rather than react. Routes are chosen with puddles in mind, schedules allow for slower commutes, and evenings turn inward. Lights come on earlier, indoor spaces feel warmer by contrast, and the outside world becomes something to move through rather than linger in.

There's a quiet value to November weather. Rain replenishes rivers, restores soil moisture, and washes the landscape clean after months of dryness. The air feels heavier, but also fresher between systems. Familiar landmarks soften under low clouds, and the region takes on a muted, reflective tone.

I think of November as the month that teaches patience without reward. There are no dramatic shifts, no clear milestones—just continuity. And in that steadiness, the Northwest finds its balance again.

Wind also finds a place in November. Fronts arrive with more energy now, and gusts rattle branches already weakened by autumn. Leaves that clung stubbornly through October finally let go, scattering across streets and trails. Power flickers occasionally reminding us that the season carries consequence as well as character.

One November windstorm stands out. Winds rose quickly after sunset, sweeping through neighborhoods and along the Portland waterfront with little warning. Trees swayed, loose debris traveled farther than expected, and the sound of wind replaced rain for several hours. The 1981 "Friday the 13th" storm ranks among the Pacific Northwest's most significant wind events of the last century. Often grouped with the region's "Big Three" windstorms, it was the strongest to strike since the Columbus Day Storm of 1962. The event was not a single storm, but a powerful one-two sequence of systems that moved through the region over the weekend of November 13–15, 1981. Wind gusts reached 85 mph in Portland. Puget Sound saw wind speeds in the 55-65 mph range.

As the month progresses, people adjust. Outings shorten. Routes become familiar. Weather forecasts regain their place in daily conversation, not out of anxiety, but habit. November doesn't ask us to endure the weather. It asks us to live alongside it.

There's a quiet honesty to the month. The sky does not pretend to be generous, and expectations fall in line accordingly. In that way, November prepares us well—for winter, and for the slower pace that comes with it.

November doesn't seek attention. It simply settles in and stays.

QUICK TAKEAWAY

November at a glance

- Frequent rain dominates

- Daylight shortens quickly

- Temperatures remain mild

- The season settles into routine

December

Quiet Return

December arrives without urgency. The Pacific Northwest is already deep into winter routines, and the weather no longer needs to announce itself. Days are short, nights are long, and the region settles into a familiar cadence of clouds, rain, and early darkness.

This is not a month of extremes. Temperatures are cool rather than cold, and snow remains uncertain west of the Cascades. Rain dominates, often steady and persistent, punctuated by brief pauses that feel almost generous. December weather doesn't surprise—it reassures through repetition.

Light becomes precious now. Even small breaks in the clouds feel meaningful, and reflections from wet pavement, windows, and water seem brighter against the dim afternoons. December teaches attention to small moments rather than grand displays.

Seattle & Portland

December weather in Seattle and Portland reflects winter's settled state. Seattle often experiences prolonged stretches of cloud cover and light rain, with only occasional breaks. Marine air keeps temperatures relatively mild, though the dampness can make days feel colder than the numbers suggest.

Portland shares much of the same pattern, with frequent rain and limited daylight shaping daily routines. Cold air occasionally slips west through the Columbia Gorge, sharpening temperatures for short periods, but sustained freezes remain uncommon. Snow is always possible, but never guaranteed.

December doesn't invite long outdoor stays, but it does encourage meaningful moments outside.

In Seattle, waterfront walks remain popular during dry breaks, especially when low clouds lift and the city's holiday lights reflect off wet pavement. Neighborhood strolls, short park visits, and shoreline viewpoints fit the season well. December is also a time when people step outside intentionally—to take in winter scenery, watch passing storms, or simply feel the air before heading back indoors.

In Portland, December outdoor activity centers on familiar routes and seasonal tradition. Walks through neighborhoods and city parks continue between rain events, and the Willamette River waterfront often draws visitors when skies briefly clear. Holiday events and displays bring people outside after dark, bundled and unhurried, even in light rain.

Across both cities, December outdoor time is shorter but purposeful. The weather doesn't discourage going outside—it simply shapes how and when it happens. December teaches that even brief moments outdoors can feel grounding, especially at the close of the year.

December dressing in the Pacific Northwest is about protection and comfort.

In both Seattle and Portland, a waterproof, insulated jacket is essential. Rain is frequent, temperatures are cool, and damp air can make conditions feel colder than the thermometer suggests. Layers underneath allow flexibility, especially when moving between indoors and outdoors.

Footwear should handle wet pavement and slick surfaces. Shoes or boots with good traction and water resistance are especially helpful during periods of heavy rain, frost, or freezing fog. In colder spells, warm socks and insulated footwear become more important.

Hats and gloves are common now, particularly during morning and evening outings. Umbrellas are widely used in Portland and useful throughout the month, while Seattle residents often continue to rely on hoods and timing their trips between showers.

December clothing favors function over fashion—items chosen to keep outings comfortable, even if brief.

If you like numbers, here's what December usually looks like.

Seattle
- Average high: 46.5 degrees F

- Average low: 37 degrees F

- Average mean temperature: 41.8 degrees F

- Average monthly rainfall: 5.55 inches

Extremes:
- Maximum: 65 degrees F, 2014

- Minimum: 9 degrees F, 1990

- Precipitation: 15.33 inches, 1933

Portland
- Average high: 46.9 degrees F

- Average low: 36.2 degrees F

- Average mean temperature: 41.6 degrees

- Average monthly rainfall: 5.77 inches

Extremes:
- Maximum: 67 degrees, 2023

- Minimum: 14 degrees F, 1968

- Precipitation: 15.24 inches, 2015

December's statistics reflect stability rather than spectacle.

WEATHER LORE

"December snow brings a fertile year."
This old saying reflects the long-held belief that winter moisture prepares the ground for what follows. In the Northwest, snow is less important than steady rain, but the sentiment holds. December weather is about preparation rather than performance.

"December rain brings January gain."
This saying reflects the idea that winter moisture does important work. In the Pacific Northwest, December rain replenishes rivers, snowpack, and soils, setting the stage

for the colder months ahead. While it can feel relentless at times, December precipitation is part of the long cycle that sustains the region.

"When December fog comes down, expect colder nights."

Fog often forms during calm, cold stretches in December, especially overnight and in the early morning hours. These foggy periods frequently signal cooling temperatures and the potential for frost or freezing conditions once skies clear. It's one of winter's quieter signals.

"A green December fills the barns."

This old saying points to a mild, wet start to winter. In the Northwest, a green December—marked by rain rather than snow at lower elevations—is common west of the Cascades. While snow lovers may be disappointed, this pattern keeps vegetation alive and reduces early-season stress on landscapes.

FROM THE WEATHER EYE

December rarely announces itself in the Pacific Northwest. It arrives quietly, with shorter days, lower light, and a steady return of rain. The season doesn't rush in—it settles, much the way January will eventually do after it.

Early December often carries a mix of moods. Some days feel mild and familiar, others turn colder without much warning. Rain becomes more persistent, and skies tend to linger low and gray. The landscape responds quickly. Lawns stay green, leaves disappear into gutters, and rivers begin their winter work.

I think of December as a month of acceptance. Expectations narrow. Outdoor plans grow simpler. The weather no longer negotiates; it establishes a pattern. And in doing so, it invites us to slow down.

Snow is never guaranteed, but it's always a possibility. When it does arrive, it often comes quietly, late at night or early in the morning, softening the familiar and briefly reshaping the routine. One December morning stands out—snow clinging to rooftops and trees, streets hushed, the city moving carefully for a few hours before rain returned to erase the evidence. These moments linger precisely because they are brief.

December weather isn't about spectacle. It's about rhythm. Rain taps steadily. Nights lengthen. The air feels heavier, but calmer. We adjust almost without noticing—layers by the door, lights turned on earlier, outings measured more thoughtfully.

By the end of the month, the cycle feels complete. The sky has settled into winter, and so have we. December doesn't ask for attention or admiration. It simply prepares the way.

And when January arrives, quietly and without cere-mony, it feels familiar—like a continuation rather than a beginning.

Another December morning in Portland stands out for how ordinary it began. Overnight rain had turned to wet snow just before dawn, coating rooftops and the West Hills while lower elevations stayed slushy. The city slowed instinctively. Traffic thinned, footsteps softened, and the familiar sounds of rain were briefly replaced by a hush that lasted only a few hours. By midday, temperatures crept up and rain returned, washing the scene away. It was a reminder of how briefly winter sometimes leaves its mark here—and how quickly the city adapts.

In Seattle, a December cold snap once settled in af-ter several days of steady rain. As temperatures dropped overnight, the rain froze on contact, glazing sidewalks and tree branches by morning. Puget Sound lay calm and steel-gray, steam rising faintly where warmer water met cold air. The city moved carefully that day, timing errands between slick stretches and waiting for temperatures to climb. By afternoon, the ice loosened, rain resumed, and the moment passed as quietly as it arrived.

I think of December as a return rather than an ending. The cycle is complete, and the quiet prepares us to begin again—with January's cold clarity waiting just ahead.

QUICK TAKEAWAY

December at a glance

- Rain dominates, snow remains uncertain

- Daylight is scarce

- Temperatures stay cool but moderate

- Winter settles in fully

Afterword

WATCHING THE SKY

Weather is never finished with us. Even after a year of observation, the sky keeps changing, and so do we.

In the Pacific Northwest, we learn early that the weather doesn't need to explain itself. It simply arrives, settles in, and moves on. Over time, we adjust—not by controlling it, but by paying attention. We dress differently. We plan loosely. We learn which days invite patience and which ones reward spontaneity.

This book wasn't written to predict what comes next. Forecasts do that well enough, even when they miss the mark. Instead, these pages are meant to reflect what experience teaches us—how the weather usually behaves, how it shapes daily life, and how it quietly connects us to the seasons whether we notice or not.

The sayings and bits of weather lore scattered through-out the year are reminders of that long relationship. Some hold truth, some miss entirely, but all of them reflect a time when people watched the sky closely because they had to. Even now, with technology at our fingertips, there's value in stepping outside, feeling the air, and making our own quiet assessment of the day ahead.

If you're new to the Northwest, I hope this book helps you settle in and feel at home under these changing skies. If you've lived here awhile, perhaps it confirms what you already know—or gives words to something you've felt but never quite named.

Either way, the invitation is the same. Keep watching. Notice the light. Listen to the rain. The weather will always have something to say, if we're willing to pay attention.

Northwest Weather Quirks

Weather in the Pacific Northwest rarely behaves the same way everywhere at once. Small changes in location—sometimes just a few miles—can lead to noticeable differences in what people experience day to day.

Where Portland's Rain Is Measured

Official rainfall totals for Portland are taken at Portland

International Airport. While that makes sense from an aviation standpoint, it doesn't always reflect what much of the metro area experiences. The airport sits along the Columbia River in a relatively open location that often receives less rain than neighborhoods farther south and west.

Areas like Lake Oswego, West Linn, and the west hills of Portland frequently see higher annual rainfall totals, especially during winter. In many ways, Lake Oswego offers a more representative snapshot of what much of the greater Portland area experiences. It's one of those quiet details longtime residents understand instinctively, even if the numbers don't always show it.

Portland's Wild Card — The Columbia Gorge

Portland's weather has a personality trait that nearby cities don't share: access to the Columbia River Gorge.

The Gorge acts like a natural wind tunnel, connecting the interior of the continent with the Willamette Valley. When high pressure builds east of the Cascades, cold, dry air can surge westward through the Gorge and spill into the Portland metro area. This is when winter temperatures drop quickly, east winds pick up, and familiar Northwest rain can turn into freezing rain or snow.

These events don't happen often, but when they do, they tend to leave an impression. Roads glaze over, trees collect ice, and the city slows to a crawl. Meanwhile, areas just a few miles away—especially south or west—may experience something entirely different.

In summer, the Gorge can work in the opposite direction. Strong easterly winds can usher in hot, dry air from the Columbia Basin, pushing temperatures higher than expected for a city so close to the Pacific Ocean.

It's a reminder that Portland's weather isn't shaped by one influence alone. The ocean matters—but so does the land, and the narrow passage that connects them.

Why Snow Shuts Portland Down So Quickly

Snow in the Portland metro area is memorable precisely because it's rare. When it does arrive, it often comes with conditions that make it far more disruptive than the snowfall itself.

Most winter events here involve cold air spilling through the Columbia Gorge, followed by moisture riding in from the Pacific. That combination favors freezing rain and ice over dry, powdery snow. Roads glaze quickly, sidewalks turn slick, and even a small amount of accumulation can create widespread problems.

Add in hills, bridges, and a city not accustomed to frequent winter storms, and it becomes clear why things slow down so fast. It's not a lack of resilience—it's a lack of repetition. Cities that see snow regularly learn to live with it. Portland treats snow as a special event, because it is.

The result is familiar: schools close, traffic stalls, and the city collectively takes a breath. And within a day or two, the rain usually returns, quietly erasing the evidence.

Seattle's Rain Gauge — Not Always the Whole Story

Seattle has a well-earned rainy reputation, and the numbers back that up: the official annual precipitation total at Seattle-Tacoma International Airport averages in the high 30-inch range (around 39+ inches per year).

But here's a nuance many people don't think about: that gauge is just one place—and one type of weather pattern—within a wide, varied urban climate.

Because Seattle's rainfall is measured at the airport, which sits south of the city proper and somewhat exposed, it can give a *different picture* of local experience than what Seattle residents actually see in their backyards.

Around the city, terrain and microclimates matter. Seattle neighborhoods built into the hills, or on the cooler, moister east side of the city, can actually get *slightly more frequent or heavier rainfall* than the airport measurement suggests, while drier pockets (like parts of Belltown or Uptown) can see less. In fact, rainfall totals across neighborhoods vary by several inches annually thanks to subtle differences in elevation and landscape.

It's a reminder that rainfall isn't uniform across the city, and that official numbers — while useful for comparison — don't always reflect what someone living in Capitol Hill, Rainier Beach, or Ballard experiences day to day.

When the Rain Can't Decide — Puget Sound Micro-climates

Rainfall around Puget Sound is anything but uniform. On some days, one neighborhood is drying out while another is getting soaked, often just a few miles away. The reason lies in a combination of geography, wind direction, and one of the Northwest's most fascinating weather features.

The Puget Sound region is shaped by water, hills, and mountain ranges that subtly steer incoming air. When winds blow in from the west, moist Pacific air moves inland and splits as it encounters the Olympic Mountains. Some of that air flows north, some south, and when those streams meet again over central Puget Sound, the air is forced upward.

That upward motion squeezes moisture out of the clouds, creating what meteorologists call the Puget Sound Convergence Zone. The result can be a narrow but intense band of rain—or even snow—while surrounding areas see little more than clouds or drizzle.

This is why places like Everett, Shoreline, or north Seattle can sometimes get hammered with steady precipitation while downtown Seattle or areas farther south stay relatively dry. It's not unusual for residents to hear reports of heavy rain nearby while their own streets remain damp but manageable.

Microclimates add another layer to the puzzle. Slight changes in elevation, proximity to water, and local terrain can shift rainfall totals noticeably over time. Hills tend to squeeze out a bit more moisture, while areas closer to open water may see lighter but more frequent drizzle.

The takeaway is simple: around Puget Sound, rainfall maps are guidelines, not guarantees. The sky doesn't follow neighborhood boundaries, and a forecast for "Seattle" may play out very differently depending on where you're standing.

It's one more reminder that Northwest weather rewards observation. Sometimes the best forecast comes from looking out the window—and listening to friends across town describe something entirely different.

A Few Miles Can Matter—

In the Northwest, geography shapes the weather in almost imperceptible but important ways. Hills, valleys, rivers, and proximity to water all play a role. It's why one neighborhood can be drying out while another is still under steady drizzle—and why local experience often matters just as much as the forecast.

A Final Word About the Weather

By the time you reach the end of this book, the weather may already have changed.

That's part of its nature, and part of its appeal. Weather doesn't wait for us to catch up, nor does it insist on being understood all at once. It moves at its own pace, offering clues rather than answers, patterns rather than promises.

For those new to the Pacific Northwest, I hope these pages have helped make sense of a climate that can feel confusing at first—one shaped by water, mountains, and subtle shifts that rarely announce themselves. For long-time residents, my hope is simpler: that something here felt familiar, or perhaps revealed itself in a slightly new way.

Weather watching, at its heart, is an act of paying attention. It asks us to notice small changes, to slow down long enough to see what's unfolding overhead and around us. Over time, those observations become stories—some remembered clearly, others quietly folded into daily life.

After many years of writing about the weather, I've come to appreciate not just the forecasts or the records, but the way weather marks our days. It shapes our routines, influences our moods, and occasionally surprises us into looking up.

Wherever you are as you read this—at home, traveling, or somewhere in between—I hope these pages have kept you company for a while. And the next time the sky looks uncertain or familiar or quietly beautiful, I hope you'll pause, take a look, and enjoy the moment.

Thanks for reading and happy weather watching!

— Pat Timm

www.ingramcontent.com/pod-product-compliance
Lightning Source LLC
Chambersburg PA
CBHW071233020426
42333CB00015B/1448

9780991020539